THE
ATHEIST'S
BIBLE

ecco
An *Imprint* of HarperCollins*Publishers*

THE
ATHEIST'S
BIBLE

an illustrious collection

of irreverent thoughts

CONCEIVED AND EDITED BY JOAN KONNER

HarperCollins books may be purchased for educational, business, or sales promotional use. For information, please write: Special Markets Department, HarperCollins Publishers, 10 East 53rd Street, New York, NY 10022.

Every effort has been made to track the quotations back to their original sources. If any errors have mistakenly occurred, they will be corrected in subsequent editions.

An extension of this copyright page appears on page 191.

FIRST EDITION

Designed by Cassandra J. Pappas

Library of Congress Cataloging-in-Publication Data
is available upon request.

ISBN: 978-0-06-134915-7
ISBN-10: 0-06-134915-1

07 08 09 10 11 ID/RRD 10 9 8 7 6 5 4 3 2 1

Editor's Note

A substantial number of the quotes in this volume
come from works of fiction. It is important to note that
authors do not necessarily agree with the words, ideas,
or opinions expressed by the characters they create.

CONTENTS

INTRODUCTION

The world would be astonished if it knew how
great a proportion of its brightest ornaments—
of those most distinguished even in popular
estimation for wisdom and virtue—are complete
skeptics in religion.

—John Stuart Mill

The reason there are so many opinions is that no
one knows the Truth.

—Joan Konner, editor

THE
ATHEIST'S
BIBLE

i

GENESIS

Perhaps our role on this planet is not to worship God—
but to create Him.

—Arthur C. Clarke

Epicurus' old questions are yet unanswered. Is [God]
willing to prevent evil, but not able? then is he impo-
tent. Is he able, but not willing? then is he malevolent. Is
he both able and willing? whence then is evil?

—David Hume

Men . . . have had the vanity to pretend that the whole creation was made for them, whilst in reality the whole creation does not suspect their existence.

—CAMILLE FLAMMARION

Fear was the gods' begetter in this world.

—PETRONIUS

We are here because one odd group of fishes had a peculiar fin anatomy that could transform into legs for terrestrial creatures; because the earth never froze entirely during an ice age; because a small and tenuous species, arising in Africa a quarter of a million years ago, has managed, so far, to survive by hook and by crook. We may yearn for a "higher" answer—but none exists.

—STEPHEN JAY GOULD

As for me, I've long resolved not to think whether man created God or God man.

—FYODOR DOSTOYEVSKY

Geology shows that fossils are of different ages. Paleontology shows a fossil sequence, the list of species representing changes through time. Taxonomy shows biological relationships among species. Evolution is the explanation that threads it all together. Creationism is the practice of squeezing one's eyes shut and wailing "Does not!"

—ANONYMOUS

If you believe that there is a God, a God that made your body, and yet you think that you can do anything with that body that's dirty, then the fault lies with the manufacturer.

—LENNY BRUCE

From the point of view of a tapeworm, man was created by God to serve the appetite of the tapeworm.

—EDWARD ABBEY

The idea of God is the sole wrong for which I cannot forgive mankind.

—MARQUIS DE SADE

Men create the gods after their own image, not only with regard to their form but with regard to their mode of life.

—ARISTOTLE

We shall tell ourselves that it would be very nice if there were a God who created the world and was a benevolent Providence, and if there were a moral order in the universe and an after-life; but it is a very striking fact that all this is exactly as we are bound to wish it to be.

—SIGMUND FREUD

Brute force crushes many plants. Yet the plants rise again. The Pyramids will not last a moment compared with the daisy. And before Buddha or Jesus spoke the nightingale sang, and long after the words of Jesus and Buddha are gone into oblivion the nightingale still will sing. Because it is neither preaching nor teaching nor commanding nor urging. It is just singing. And in the beginning was not a Word, but a chirrup.

—D. H. Lawrence

At the beginning there was the Word—at the end just the Cliché.

—Stanislaw Jerzy Lec

All great truths begin as blasphemies.

—George Bernard Shaw

God is a word to express, not our ideas, but the want of them.

—John Stuart Mill

Atheists are often charged with blasphemy, but it is a crime they cannot commit.... When the Atheist examines, denounces, or satirizes the gods, he is not dealing with persons but with ideas. He is incapable of insulting God, for he does not admit the existence of any such being.

—G. W. Foote

I have too much respect for the idea of God to hold Him responsible for such an absurd world.

—GEORGES DUHAMEL

Man is, and always has been, a maker of gods.

—JOHN BURROUGHS

The whole religious complexion of the modern world is due to the absence from Jerusalem of a Lunatic Asylum.

—HAVELOCK ELLIS

Whatever we cannot easily understand we call God; this saves much wear and tear on the brain tissues.

—EDWARD ABBEY

Since it is obviously inconceivable that all religions can be right, the most reasonable conclusion is that they are all wrong.

—CHRISTOPHER HITCHENS

Heresy is only another word for freedom of thought.

—Graham Greene

Faith is doubt.

—Emily Dickinson

It is historically true that a large proportion of infidels in all ages have been persons of distinguished integrity and honour.

—John Stuart Mill

All thinking men are atheists.

—Ernest Hemingway

All children are atheists—they have no idea of God.

—Baron d'Holbach

It is, I think, an error to believe that there is any need of religion to make life seem worth living.

—Sinclair Lewis

I count religion but a childish toy,
And hold there is no sin but ignorance.

—Christopher Marlowe

If there is a God, a caring God, then we have to figure he's done an extraordinary job of making a very cruel world.

—Dave Matthews

theology—An effort to explain the unknowable by putting it into terms of the not worth knowing.

—H. L. Mencken

The universe runs itself, and the eternal laws inherent in Nature suffice, without any first cause or prime mover.

—Marquis de Sade

Martyrdom ... is the only way in which a man can become famous without ability.

—George Bernard Shaw

I contend that we are both atheists. I just believe in one fewer god than you do. When you understand why you dismiss all the other possible gods, you will understand why I dismiss yours.

—Stephen Henry Roberts

In some awful, strange, paradoxical way, atheists tend to take religion more seriously than the practitioners.

—Jonathan Miller

iii

THE GOSPEL

For it is with the mysteries of our Religion, as with whole-some pills for the sick, which swallowed whole, have the virtue to cure, but chewed, are for the most part cast up again without effect.

—THOMAS HOBBES

Marx was wrong. Religion is not the opiate of the people. Opium suggests something soporific, numbing, dulling. Too often religion has been an aphrodisiac for horror, a Benzedrine for bestiality. At its best it has lifted spirits and raised spires. At its worst it has turned entire civilizations into cemeteries.

—PHILLIP ADAMS

[Religion is] the fashionable substitute for Belief.

—Oscar Wilde

If I were personally to define religion I would say that it is a bandage that man has invented to protect a soul made bloody by circumstance.

—Theodore Dreiser

The Church is the world's great lost and found department.

—Robert L. Short

Religion is what keeps the poor from murdering the rich.

—Napoléon Bonaparte

In church, sacred music would make believers of us all— but preachers can be counted on to restore the balance.

—Mignon McLaughlin

Emotional excitement reaches men through tea, tobacco, opium, whiskey and religion.

—George Bernard Shaw

What mean and cruel things men can do for the love of God.

—W. Somerset Maugham

Men never do evil so completely and cheerfully as when they do it from religious conviction.

—Blaise Pascal

Generally speaking, the errors in religion are dangerous; those in philosophy only ridiculous.

—David Hume

Well, I've often thought that the Bible should have a disclaimer at the front saying, "This is fiction." I mean, walking on water? I mean, it takes an act of faith.

—Ian McKellen

Religion hides many mischiefs from suspicion.

—Christopher Marlowe

The devil can cite Scripture for his purpose.

—William Shakespeare

Every dogma has its day.

—Israel Zangwill

Most sermons sound to me like commercials—but I can't make out whether God is the Sponsor or the Product.

—Mignon McLaughlin

The religions of mankind must be classed among the mass-delusions of this kind. No one, needless to say, who shares a delusion ever recognizes it as such.

—Sigmund Freud

The Bible, you know, is rather a disappointment: it has never done for humanity what it should have done.

—Christopher Morley

iv

BOOK OF ENLIGHTENMENT

There was a time when religion ruled the world. It is known as the Dark Ages.

—Ruth Hurmence Green

FAITH: The effort to believe that which your common-sense tells you is not true.

—Elbert Hubbard

An actually existent fly is more important than a possibly existent angel.

—Ralph Waldo Emerson

Where it is a duty to worship the sun, it is pretty sure to be a crime to examine the laws of heat.

—John Morley

The beginning of wisdom is the awareness that there is insufficient evidence that a god or gods have created us and the recognition that we are responsible in part for our own destiny.

—Paul Kurtz

Now faith is the substance of things hoped for, the evidence of things not seen.

—Saint Paul (Hebrews 11:1)

Blind faith is an ironic gift to return to the Creator of human intelligence.

—Anonymous

How many observe Christ's Birth-day! How few,
his Precepts! O! 'tis easier to keep Holidays
than Commandments.

—BENJAMIN FRANKLIN

[Religious thought is] an attempt to find an out where
there is no door.

—ALBERT EINSTEIN

Indeed, speaking generally, religion is the *chef d'oeuvre* of
training, namely training the ability to think. . . . There
is no absurdity, however palpable, which cannot be
firmly implanted in the minds of all, if only one begins
to inculcate it before the early age of six by constantly
repeating it to them with an air of great solemnity. For
the training of man, like that of animals, is completely
successful only at an early age.

—ARTHUR SCHOPENHAUER

If God wants us to do a thing he should make his wishes sufficiently clear. Sensible people will wait till he has done this before paying much attention to him.

—Samuel Butler

I cannot believe in a God who has neither humor nor common sense.

—W. Somerset Maugham

QUESTION: How do you know you're God?
ANSWER: Simple. When I pray to him, I find I'm talking to myself.

—Peter O'Toole

We owe almost all our knowledge, not to those who have agreed, but to those who have differed.

—Charles Caleb Colton

Human life has no meaning independent of itself. . . .
The meaning of life is what we choose to give it.

—Paul Kurtz

Mankind has been punished long and heavily for having
created its gods; nothing but pain and persecution have
been man's lot since gods began. There is but one way
out of this blunder: Man must break his fetters which
have chained him to the gates of heaven and hell, so that
he can begin to fashion out of his reawakened and illu-
mined consciousness a new world upon earth.

—Emma Goldman

We don't know one millionth of one percent about anything!

—Thomas Edison

The church says the earth is flat, but I know that it is
round, for I have seen the shadow on the moon, and I
have more faith in a shadow than in the church.

—Ferdinand Magellan

The present age ... prefers the sign to the thing signified, the copy to the original, fancy to reality, the appearance to the essence ... for in these days *illusion* only is *sacred, truth profane*.

—Ludwig Feuerbach

Three-quarters of the American population literally believes in religious miracles. The numbers who believe in the devil, in resurrection, God does this and that— astonishing. These are numbers that you have nowhere in the industrial world.

—Noam Chomsky

They say that God is everywhere, and yet we always think of Him as somewhat of a recluse.

—Emily Dickinson

Men will wrangle for religion; write for it; fight for it; die for it; any thing but—*live for it*.

—Charles Caleb Colton

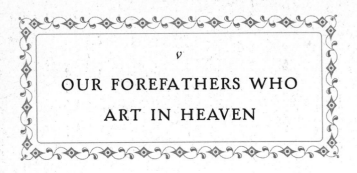

v
OUR FOREFATHERS WHO
ART IN HEAVEN

Mystery [the divinity of Jesus Christ] is made a convenient Cover for absurdity.

—JOHN ADAMS

But it does me no injury for my neighbour to say there are twenty gods, or no god. It neither picks my pocket nor breaks my leg.

—THOMAS JEFFERSON

The Bible is not my book nor Christianity my profession.

—ABRAHAM LINCOLN

Every other sect suppos[es] itself in possession of all truth, and that those who differ are so far in the wrong; like a man traveling in foggy weather, those at some distance before him on the road he sees wrapped up in the fog, as well as those behind him, and also the people in the fields on each side, but near him all appears clear, tho' in truth he is as much in the fog as any of them.

—BENJAMIN FRANKLIN

May it not suffice for me to say . . . that, of course, like every other man of intelligence and education, I do believe in organic evolution. It surprises me that at this late date such questions should be raised.

—WOODROW WILSON (1922)

Religion is excellent stuff for keeping common people quiet.

—Napoléon Bonaparte

The world is my country; to do good my religion.

—Thomas Paine

I prayed for freedom twenty years but received no answer until I prayed with my legs.

—Frederick Douglass

If I should go out of church whenever I hear a false sentiment, I could never stay there five minutes.

—Ralph Waldo Emerson

Do you think I am superstitious? I am a super-atheist.

—Mohandas K. Gandhi

I want nothing to do with any religion concerned with keeping the masses satisfied to live in hunger, filth and ignorance. I want nothing to do with any order, religious or otherwise, which does not teach people that they are capable of becoming happier and more civilized, on this earth, capable of becoming true *man*, master of his fate and captain of his soul.

—Jawaharlal Nehru

Every step which the intelligence of Europe has taken has been in spite of the clerical party.

—Victor Hugo

It is ridiculous to think that a supreme being—whatever it is—cares about human affairs. Don't we believe that it would be defiled by so gloomy and complex a responsibility?

—Pliny the Elder

Religious hatreds ought not be propagated at all, but certainly not on a tax-exempt basis.

—James A. Michener

Organized religion is making Christianity political rather than making politics Christian.

—Laurens van der Post

The number, the industry, and the morality of the Priesthood, & the devotion of the people have been manifestly increased by the total separation of the Church from the State.

—James Madison

And of all Plagues with which Mankind are Curst,
Ecclesiastick Tyranny's the worse.

—Daniel Defoe

Who does not see that the same authority which can establish Christianity, in exclusion of all other Religions, may establish with the same ease any particular sect of Christians, in exclusion of all other Sects?

—James Madison

A tyrant . . . should always show a particular zeal in the cult of the gods. People are less afraid of being treated unjustly by those of this sort, that is if they think that the ruler is god-fearing and pays some regard to the gods; and they are less ready to conspire against him, if they feel that the gods themselves are his friends.

—ARISTOTLE

I'm frankly sick and tired of the political preachers across this country telling me as a citizen that if I want to be a moral person, I must believe in "A," "B," "C," and "D." Just who do they think they are? And from where do they presume to claim the right to dictate their moral beliefs to me? And I am even more angry as a legislator who must endure the threats of every religious group who thinks it has some God-granted right to control my every roll call in the Senate. I am warning them today: I will fight them every step of the way if they try to dictate their moral convictions to all Americans in the name of "conservatism."

—BARRY GOLDWATER

The need for religion will end when man becomes sensible enough to govern himself.

—FRANCISCO FERRER GUARDIA

vi

BOOK OF REASON

Question with boldness even the existence of a god; because, if there be one, he must more approve of the homage of reason, than that of blindfolded fear.

—THOMAS JEFFERSON

The fact that a believer is happier than a skeptic is no more to the point than the fact that a drunken man is happier than a sober one.

—GEORGE BERNARD SHAW

It is vain to ask of the gods what a man is capable of supplying for himself.

—Epicurus

Theology is a thing of unreason altogether, an edifice of assumptions and dreams, a superstructure without a substructure.

—Ambrose Bierce

Gullibility and credulity are considered undesirable qualities in every department of human life—except religion. . . . Why are we praised by godly men for surrendering our "godly gift" of reason when we cross their mental thresholds?

—Christopher Hitchens

This story of the redemption will not stand examination. That man should redeem himself from the sin of eating an apple, by committing a murder on Jesus Christ, is the strangest system of religion ever set up.

—Thomas Paine

You can't convince a believer of anything; for their belief is not based on evidence, it's based on a deep-seated need to believe.

—Carl Sagan

There is no God; it is as clear as the sun and as evident as the day that there is no God, and still more, that there *can* be none.

—Ludwig Feuerbach

I get letters constantly from people saying, "Oh, God will look after it." But He never has in the past, I don't know why they think He will in the future.

—Bertrand Russell

There's never been anything, however absurd, that myriads of people weren't prepared to believe, often so passionately that they'd fight to the death rather than abandon their illusions. To me, that's a good operational definition of insanity.

—Arthur C. Clarke

All religions, with their gods, their demigods, and their prophets, their messiahs and their saints, were created by the credulous fancy of men who had not attained the full development and full possession of their faculties.

—MIKHAIL A. BAKUNIN

The most formidable weapon against errors of every kind is Reason. I have never used any other, and I trust I never shall.

—THOMAS PAINE

What is wanted is not the will to believe, but the wish to find out, which is the exact opposite.

—BERTRAND RUSSELL

When lip service to some mysterious deity permits bestiality on Wednesday and absolution on Sunday—cash me out.

—FRANK SINATRA

All religions promise a reward beyond this life in eternity for excellences of the *will* or of the heart, but none for excellences of the head, of the understanding.

—ARTHUR SCHOPENHAUER

The Christian religion not only was at first attended with miracles, but even at this day cannot be believed by any reasonable person without one.

—DAVID HUME

Without cultural sanction, most or all of our religious beliefs and rituals would fall into the domain of mental disturbance.

—JOHN F. SCHUMAKER

Do not let yourself be deceived: great intellects are skeptical.

—FRIEDRICH NIETZSCHE

If this is your God, he's not very impressive. He has so many psychological problems; he's so insecure. He demands worship every seven days. He goes out and creates faulty humans and then blames them for his own mistakes. He's a pretty poor excuse for a Supreme Being.

—GENE RODDENBERRY

I find the whole business of religion profoundly interesting. But it does mystify me that otherwise intelligent people take it seriously.

—DOUGLAS ADAMS

If we think that this search for God is a vain search, and that there is no reality to be discovered ... then the history of religion becomes a study of the aberrations of the human mind.

—CYRIL BAILEY

An atheist ... is a man who destroys chimeras harmful to the human race, in order to lead men back to nature, to experience, and to reason.

—BARON D'HOLBACH

Perhaps the whole root of our trouble, the human trouble, is that we will sacrifice all the beauty of our lives, will imprison ourselves in totems, taboos, crosses, blood sacrifices, steeples, mosques, races, armies, flags, nations, in order to deny the fact of death, which is the only fact we have.

—JAMES BALDWIN

I would believe any religion that could prove it had existed since the beginning of the world. But when I see Socrates, Plato, Moses, and Mohammed I do not think there is such a one. All religions owe their origin to man.

—NAPOLÉON BONAPARTE

Indeed, when religious people quarrel about religion, or hungry people about their victuals, it looks as if they had not much of either among them.

—BENJAMIN FRANKLIN

I have seldom met an intelligent person whose views were not narrowed and distorted by religion.

—James Buchanan

Those who take refuge behind theological barbed wire fences, quite often wish they could have more freedom of thought, but fear the change to the great ocean of scientific truth as they would a cold bath plunge.

—Luther Burbank

A *material* resurrection seems strange, and even absurd, except for purposes of punishment; and all punishment, which is to *revenge* rather than *correct*, must be *morally wrong*. And *when* the *World is at an end*, what moral or warning purpose *can* eternal tortures answer?

—Lord Byron

I shall not, as far as I am concerned, try to pass myself off as a Christian in your presence.* I share with you the same revulsion from evil. But I do not share your hope, and I continue to struggle against this universe in which children suffer and die.

—ALBERT CAMUS

All religions have this in common, that they are an outrage to common sense, for they are pieced together out of a variety of elements, some of which seem so unworthy, sordid, and at odds with man's reason that any strong and vigorous intelligence laughs at them.

—PIERRE CHARRON

The idea that He would take his attention away from the universe in order to give me a bicycle with three speeds is just so unlikely that I can't go along with it.

—QUENTIN CRISP

* Attributed to statements Camus made at the Dominican Monastery of Latour-Maubourg in 1948.

I cannot persuade myself that a beneficent and omnipotent God would have designedly created . . . that a cat should play with mice.

—Charles Darwin

> Faith is a fine invention
> For gentlemen who see;
> But microscopes are prudent
> In an emergency!
> —Emily Dickinson

Faith is believing things by definition, which are not justified by reason. If it were justified by reason, it wouldn't be faith.

—Colin McGinn

Dogmas of every kind put assertion in the place of reason and give rise to more contention, bitterness, and want of charity than any other influence in human affairs.

—Sir Arthur Conan Doyle

If God listened to the prayers of men, all men would quickly have perished; for they are for ever praying for evil against one another.

—Epicurus

Parsons always seem to be specially horrified about things like sunbathing and naked bodies. They don't mind poverty and misery and cruelty to animals nearly as much.

—Susan Ertz

In the long run nothing can withstand reason and experience, and the contradiction which religion offers to both is all too palpable.

—Sigmund Freud

I recall the story of the philosopher and the theologian. The two were engaged in disputation and the theologian used the old quip about a philosopher being like a blind man, in a dark room, looking for a black cat—which wasn't there. "That may be," said the philosopher; "but a theologian would have found it."

—Julian Huxley

The existence of a world without God seems to me less absurd than the presence of a God, existing in all his perfection, creating an imperfect man in order to make him run the risk of Hell.

—Armand Salacrou

I did not see why the schoolmaster should be taxed to support the priest, and not the priest the schoolmaster.

—Henry David Thoreau

The difference between religions and cults is determined by how much real estate is owned.

—Frank Zappa

It is an insult to God to believe in God. For on the one hand it is to suppose that he has perpetrated acts of incalculable cruelty. On the other hand, it is to suppose that he has perversely given his human creatures an instrument—their intellect—which must inevitably lead them, if they are dispassionate and honest, to deny his existence. It is tempting to conclude that if he exists, it is the atheists and agnostics that he loves best, among those with any pretensions to education. For they are the ones who have taken him most seriously.

—Galen Strawson

The Way to see by *Faith* is to shut the Eye of *Reason*.

—Benjamin Franklin

vii

SCIENTOSOPHY*

[Science] is the true theology.

—THOMAS PAINE

The Religion that is afraid of science dishonours God & commits suicide.

—RALPH WALDO EMERSON

* "Scientosophy" is an invented word, implying that science is a belief system, as much as any other. The chapter includes quotes from those who believe that science will or does reveal, beyond facts, The Truth, demonstrating that believers in science can be as dogmatic and faithful to their beliefs as some religions and religious believers are to theirs.

There is no personal salvation, there is no national salvation, except through science.

—Luther Burbank

My mind is incapable of conceiving such a thing as a soul. I may be in error, and man may have a soul; but I simply do not believe it.

—Thomas Edison

Once miracles are admitted, every scientific explanation is out of the question.

—Johannes Kepler

Anything that we scientists can do to weaken the hold of religion should be done and may in the end be our greatest contribution to civilization.

—Steven Weinberg

If we are going to teach "creation science" . . . as an alternative to evolution, then we should also teach the stork theory as an alternative to biological reproduction.

—JUDITH HAYES

The most incomprehensible thing about the world is that it is comprehensible.

—ALBERT EINSTEIN

There is a very intimate connection between hypnotic phenomena and religion.

—HAVELOCK ELLIS

Science . . . has opened our eyes to the vastness of the universe and given us light, truth and freedom from fear where once was darkness, ignorance and superstition.

—LUTHER BURBANK

And certainly we should take care not to make the intellect our god; it has, of course, powerful muscles, but no personality.

—Albert Einstein

It is wrong always, everywhere, and for anyone, to believe anything upon insufficient evidence.

—W. K. Clifford

None of the miracles with which ancient histories are filled, occurred under scientific conditions. Observation never once contradicted, teaches us that miracles occur only in periods and countries in which they are believed in and before persons disposed to believe in them.

—Ernest Renan

Religions are conclusions for which the facts of nature supply no major premises.

—Ambrose Bierce

In science it often happens that scientists say, "You know that's a really good argument; my position is mistaken," and then they actually change their minds and you never hear that old view from them again. They really do it. It doesn't happen as often as it should, because scientists are human and change is sometimes painful. But it happens every day. I cannot recall the last time something like that has happened in politics or religion.

—Carl Sagan

Science can destroy religion by ignoring it as well as by disproving its tenets. No one ever demonstrated, so far as I am aware, the nonexistence of Zeus or Thor, but they have few followers now.

—Arthur C. Clarke

One of the most frightening things in the Western world, and this country in particular, is the number of people who believe in things that are scientifically false. If someone tells me that the earth is less than 10,000 years old, in my opinion he should see a psychiatrist.

—Francis Crick

It seems to me that the idea of a personal God is an anthropological concept which I cannot take seriously. I feel also not able to imagine some will or goal outside the human sphere.

—Albert Einstein

Our knowledge of the historical worth of certain religious doctrines increases our respect for them, but does not invalidate our proposal that they should cease to be put forward as the reasons for the precepts of civilization. . . . We may now argue that the time has probably come . . . for replacing the effects of repression by the results of the rational operation of the intellect.

—Sigmund Freud

Some people think of God as an outsized, light-skinned male with a long white beard, sitting on a throne somewhere up there in the sky, busily tallying the fall of every sparrow. Others—for example, Baruch Spinoza and Albert Einstein—considered God to be essentially the sum total of the physical laws which describe the universe. I do not know of any compelling evidence for anthropomorphic patriarchs controlling human destiny from some hidden celestial vantage point, but it would be madness to deny the existence of physical laws.

—CARL SAGAN

Faith is the great cop-out, the great excuse to evade the need to think and evaluate evidence. Faith is belief in spite of, even perhaps because of, the lack of evidence.

—RICHARD DAWKINS

This century will be called Darwin's century. He was one of the greatest men who ever touched this globe. He has explained more of the phenomena of life than all of the religious teachers. . . . His doctrine of evolution, his doctrine of the survival of the fittest, his doctrine of the origin of species, has removed in every thinking mind the last vestige of orthodox Christianity. He has not only stated, but he has demonstrated . . . that the Bible is a book written by ignorance—at the instigation of fear.

—ROBERT G. INGERSOLL

Science has done more for the development of Western civilization in one hundred years than Christianity did in eighteen hundred.

—JOHN BURROUGHS

When the masses become better informed about science, they will feel less need for help from supernatural Higher Powers.

—FRANCISCO FERRER GUARDIA

Gods are fragile things; they may be killed by a whiff of science or a dose of common sense.

—CHAPMAN COHEN

Religion is an illusion and it derives its strength from its readiness to fit in with our instinctual wishful impulses.

—SIGMUND FREUD

Don't believe without evidence. Treat things divine with marked respect—don't have anything to do with them.

—AMBROSE BIERCE

Nature and Nature's laws lay hid in night;
God said, "Let Newton be!" and all was light.

—ALEXANDER POPE

Not to be absolutely certain is, I think, one of the essential things in rationality.

—BERTRAND RUSSELL

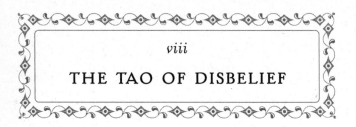

viii

THE TAO OF DISBELIEF

It is clear that he who adheres to the Church's teaching as to an infallible rule, assents to all points of that teaching.

—Saint Thomas Aquinas

I cannot be angry at God, in whom I do not believe.

—Simone de Beauvoir

If there were not God, there would be no atheists.

—G. K. Chesterton

Religion provides the solace for the turmoil that it creates.

—Byron Danelius

I am a deeply religious nonbeliever. . . . This is a somewhat new kind of religion.

—Albert Einstein

There can be no Creator, simply because his grief at the fate of his creation would be inconceivable and unendurable.

—Elias Canetti

Those who love God are not always the friends of their fellow-men.

—Robert G. Ingersoll

Don't look for God where He is needed most; if you didn't bring Him there, He isn't there.

—Mignon McLaughlin

Only the atheist realizes how morally objectionable it is for survivors of a catastrophe to believe themselves spared by a loving God while this same God drowned infants in their cribs.

—Sam Harris

To be absolutely certain about something, one must know everything or nothing about it.

—Olin Miller

I'm an atheist. There's a lot about the Catholic Church I don't approve of, simply because they don't approve of me.

—Ian McKellen

Some like to understand what they believe in. Others like to believe in what they understand.

—Stanislaw Jerzy Lec

Men make themselves believe that they believe.

—Michel Eyquem de Montaigne

Thanks be to God, I am still an atheist.

—Luis Buñuel

It has even been said that the highest praise of God is to be found in the denial of Him by the atheist, who considers creation to be perfect enough to dispense with a Creator.

—Marcel Proust

God bless those pagans.

—Homer Simpson

We must believe in free will; we have no choice.

—Isaac Bashevis Singer

I'm a born-again atheist.

—Gore Vidal

What can be said at all can be said clearly; and whereof one cannot speak thereof one must be silent.

—Ludwig Wittgenstein

ix

BOOK OF QUESTIONS

If he is infinitely good, what reason should we have to fear him? If he is infinitely wise, why should we have doubts concerning our future? If he knows all, why warn him of our needs and fatigue him with our prayers? If he is everywhere, why erect temples to him?

—PERCY BYSSHE SHELLEY

When I was in Ireland performing my one-man show . . . I told the audience I was an atheist and a woman got up and said, "Yes, but is it the God of the Catholics or the God of the Protestants in whom you do not believe?"

—QUENTIN CRISP

Isn't it enough to see that a garden is beautiful without having to believe that there are fairies at the bottom of it too?

—Douglas Adams

The biblical account of Noah's Ark and the Flood is perhaps the most implausible story for fundamentalists to defend. Where, for example, while loading his ark, did Noah find penguins and polar bears in Palestine?

—Judith Hayes

On his deathbed, Henry David Thoreau was asked by his aunt: "Henry, have you made your peace with God?" Thoreau answered, "I did not know we had ever quarrelled."

—Edward Waldo Emerson

Somewhere, and I can't find where, I read about an Eskimo hunter who asked the local missionary priest, "If I did not know about God and sin, would I go to hell?" "No," said the priest, "not if you did not know." "Then why," asked the Eskimo earnestly, "did you tell me?"

—Annie Dillard

If absolute power corrupts absolutely, where does that leave God?

—George Daacon

We have fools in all sects, and impostors in most; why should I believe mysteries no one can understand, because written by men who chose to mistake madness for inspiration and style themselves Evangelicals?

—Lord Byron

So if, as I have shown, gods do not have a human appearance, and if, as you firmly believe, they are unlike anything in the heavens, why do you hesitate to deny that they exist? You lose your nerve, and it is wise of you to do so, though your fear on this account is not of the people but of the gods themselves.

—Cicero

It is told that the great Angelo, in decorating a church, painted some angels wearing sandals. A cardinal looking at the picture said to the artist: "Whoever saw angels with sandals?" Angelo answered with another question: "Whoever saw an angel barefooted?"

—Robert G. Ingersoll

I am one of the lingering *bad* ones, and so do *I* slink away, and pause, and ponder, and ponder, and pause, and do work without knowing why—not surely for *this* brief world, and more sure it is not for Heaven.

—Emily Dickinson

Subtract from the New Testament the miraculous and highly improbable, and what will be the remainder?

—George Eliot

Why should I fear death? If I am, death is not. If death is, I am not. Why should I fear that which cannot exist when I do?

—Epicurus

We enter church, and we have to say, "We have erred and strayed from Thy ways like lost sheep," when what we want to say is, "Why are we made to err and stray like lost sheep?"

—Thomas Hardy

question: I just wanted to know if you believe in God.
answer: No, but I don't disbelieve in her either.

—Arthur C. Clarke

Must then a Christ perish in torment in every age to save those that have no imagination?

—George Bernard Shaw

I'm not a bad guy! I work hard, and I love my kids. So why should I spend half my Sunday hearing about how I'm going to Hell?

—Homer Simpson

Christ and Moses standing in the back of St. Pat's, looking around. Confused, Christ is, at the grandeur of the interior, the baroque interior, the rococo baroque interior. Because his route took him through Spanish Harlem, and he was wondering what the hell fifty Puerto Ricans were doing living in one room when that stained glass window is worth ten G's a square foot?

—Lenny Bruce

Philosophy is questions that may never be answered. Religion is answers that may never be questioned.

—Anonymous

x
REVELATIONS

If triangles made a God, they would give him three sides.

—Montesquieu

One thing I have no worry about is whether God exists. But it has occurred to me that God has Alzheimer's and has forgotten we exist.

—Jane Wagner

TRAVELER: God has been mighty good to your fields, Mr. Farmer.

FARMER: You should have seen how he treated them when I wasn't around.

—ANONYMOUS

When I was a young boy, my father taught me that to be a good Catholic, I had to confess at church if I ever had impure thoughts about a girl. That very evening I had to rush to confess my sin. And the next night, and the next. After a week, I decided religion wasn't for me.

—FIDEL CASTRO

From time to time, as we all know, a sect appears in our midst announcing that the world will very soon come to an end. Generally, by some slight confusion or miscalculation, it is the sect that comes to an end.

—G. K. CHESTERTON

I turned to speak to God
About the world's despair;
But to make bad matters worse
I found God wasn't there.

—Robert Frost

Every event, or appearance, or accident which seems to deviate from the ordinary course of nature has been rashly ascribed to the immediate action of the Deity.

—Edward Gibbon

God, Satan, Paradise and Hell all vanished one day in my fifteenth year, when I quite abruptly lost my faith. I recall it vividly. I was at school in England by then. The moment of awakening happened, in fact, during a Latin lesson, and afterwards, to prove my new-found atheism, I bought myself a rather tasteless ham sandwich, and so partook for the first time of the forbidden flesh of the swine. No thunderbolt arrived to strike me down. I remember feeling that my survival confirmed the correctness of my new position.

—Salman Rushdie

I believe in God; I just don't trust anyone who works for him.

—Anonymous

The three great apostles of practical atheism, that make converts without persecuting, and retain them without preaching, are Wealth, Health, and Power.

—Charles Caleb Colton

From the first moment I looked into that horror on September 11th, into that fireball, into that explosion of horror, I knew it. I knew it before anything was said about those who did it or why. I recognized an old companion. I recognized religion.

—Monsignor Lorenzo Albacete

All gods are dead except the god of war.

—Eldridge Cleaver

Don't wait for the Last Judgment. It takes place every day.

—Albert Camus

xi

THE GOOD BOOK

Properly read, the Bible is the most potent force for atheism ever conceived.

—Isaac Asimov

Our Bible reveals to us the character of our God with minute and remorseless exactness. . . . It is perhaps the most damnatory biography that exists in print anywhere.

—Mark Twain

You believe in a book which has talking animals, wizards, witches, demons, sticks turning into snakes, food falling from the sky, people walking on water, and all sorts of magical, absurd, and primitive stories; and you say that *I* am the one who is mentally ill?

—Dan Barker

In a heated controversy over the wisdom of giving the Bible to slaves, [Frederick Douglass] asserted that it would be "infinitely better to send them a pocket compass and a pistol."

—Benjamin Quarles

The Bible. That is what fools have written, what imbeciles command, what rogues teach, and young children are made to learn by heart.

—Voltaire

I am convinced now that children should not be subjected to the frightfulness of the Christian religion. . . .
If the concept of a father who plots to have his own son put to death is presented to children as beautiful and as worthy of society's admiration, what types of human behavior can be presented to them as *reprehensible*?

—RUTH HURMENCE GREEN

The continually progressive change to which the meaning of words is subject, the want of a universal language which renders translation necessary, the errors to which translations are again subject, the mistakes of copyists and printers, together with the possibility of willful alteration, are of themselves evidences that human language, whether in speech or in print, cannot be the vehicle of the Word of God.

—THOMAS PAINE

It seems that if our species ever eradicates itself through war, it will not be because it was written in the stars but because it was written in our books; it is what we do with words like "God" and "paradise" and "sin" in the present that will determine our future.

—SAM HARRIS

It's the generally accepted privilege of theologians to stretch the heavens, that is, the Scriptures, like tanners with a hide.

—Desiderius Erasmus

The Gospels actually taught this: *Before you kill somebody, make absolutely sure he isn't well connected.*

—Kurt Vonnegut, Jr.

There is much in the Bible against which every instinct of my being rebels, so much that I regret the necessity which has compelled me to read it through from beginning to end. I do not think that the knowledge which I have gained of its history and sources compensates me for the unpleasant details it has forced upon my attention.

—Helen Keller

In religion,
What damned error but some sober brow
Will bless it and approve it with a text,
Hiding the grossness with fair ornament?
—WILLIAM SHAKESPEARE

The Bible, itself the ultimate curse, is an in-depth pro-file of the divine spleen.

—RUTH HURMENCE GREEN

As to the book called the Bible, it is blasphemy to call it the word of God. It is a book of lies and contradictions, and a history of bad times and bad men. There are but a few good characters in the whole book.

—THOMAS PAINE

When I think of all the harm [the Bible] has done, I despair of ever writing anything to equal it.

—OSCAR WILDE

The Old Testament is responsible for more atheism, agnosticism, disbelief—call it what you will—than any book ever written; it has emptied more churches than all the counterattractions of cinema, motor bicycle and golf course.

—A. A. Milne

"The Good Book"—one of the most remarkable euphemisms ever coined.

—Ashley Montagu

xii

THE BRIMSTONE CHRONICLES *

Civilization will not attain to its perfection, until the last stone from the last church falls on the last priest!

—Émile Zola

To aim to convert a man by miracles is a profanation of the soul.

—Ralph Waldo Emerson

* "The Brimstone Chronicles" derives its name from preachers who scare their congregations into belief in and submission to God through fear of hell and damnation. The quotes in this chapter demonstrate that Atheists can be as intemperate, unreasonable and extreme as fire-and-brimstone preachers.

Religion has done more to bust-up humanity than any-thing.

—Whoopi Goldberg

Examine the religious principles which have, in fact, prevailed in the world. You will scarcely be persuaded that they are any thing but sick men's dreams.

—David Hume

There is one notable thing about our Christianity . . . it is still a hundred times better than the Christianity of the Bible, with its prodigious crime—the invention of Hell. Measured by our Christianity of today, bad as it is, hypocritical as it is, empty and hollow as it is, neither the Deity nor His Son is a Christian, nor qualified for that moderately high place.

—Mark Twain

God is the immemorial refuge of the incompetent, the helpless, the miserable. They find not only sanctuary in His arms, but also a kind of superiority, soothing to their macerated egos: He will set them above their betters.

—H. L. MENCKEN

I find it necessary to wash my hands after I have come into contact with religious people.

—FRIEDRICH NIETZSCHE

All your Western theologies, the whole mythology of them, are based on the concept of God as a *senile delinquent.*

—TENNESSEE WILLIAMS

Christianity is the most ridiculous, the most absurd and bloody religion that has ever infected the world.

—VOLTAIRE

xiii

PROVERBS

Religion is regarded by the common people as true, by the philosophers as false, and by the rulers as useful.

—Seneca the Younger

It is an heretic that makes the fire,
Not she which burns in't.

—William Shakespeare

But no public man . . . ever believes that the Bible means what it says: he is always convinced that it says what he means.

—George Bernard Shaw

I have heard an atheist defined as a man who had no invisible means of support.

—John Buchan

Give a man a fish, and you'll feed him for a day; give him a religion, and he'll starve to death while praying for a fish.

—Anonymous

A simple man believes anything,
but a prudent man gives thought to his steps.

—Proverbs, 14:15

One's convictions should be proportional to one's evidence.

—Sam Harris

To know a person's religion we need not listen to his profession of faith but must find his brand of intolerance.

—Eric Hoffer

Art raises its head where religions decline.

—Friedrich Nietzsche

As the caterpillar chooses the fairest leaves to lay her eggs on, so the priest lays his curse on the fairest joys.

—William Blake

A faith which cannot survive collision with the truth is not worth many regrets.

—Arthur C. Clarke

Religion is for people who are afraid to go to hell, whereas spirituality is for people like me who have been there.

—Dave Mustaine

The religion of one age is the literary entertainment of the next.

—RALPH WALDO EMERSON

For every credibility gap, there is a gullibility fill.

—RICHARD CLOPTON

He, who begins by loving Christianity better than Truth, will proceed by loving his own Sect or Church better than Christianity, and end in loving himself better than all.

—SAMUEL TAYLOR COLERIDGE

A dream is a scripture, and many scriptures are nothing but dreams.

—UMBERTO ECO

As men's prayers are a disease of the will, so are their creeds a disease of the intellect.

—RALPH WALDO EMERSON

A myth is a religion in which no one any longer believes.

—James K. Feibleman

Lighthouses are more helpful than churches.

—Benjamin Franklin

There is in every village a torch: the schoolmaster—and an extinguisher: the parson.

—Victor Hugo

Praying is like a rocking chair—it'll give you something to do, but it won't get you anywhere.

—Gypsy Rose Lee

Nothing is so firmly believed as what is least known.

—Michel Eyquem de Montaigne

Absolute faith corrupts as absolutely as absolute power.

—Eric Hoffer

All religions are equally sublime to the ignorant, useful to the politician, and ridiculous to the philosopher.

—Lucretius

Two hands working can do more than a thousand clasped in prayer.

—Anonymous

xiv

BOOK OF COMMON VIRTUE

The greatest tragedy in mankind's entire history may be the hijacking of morality by religion.

—ARTHUR C. CLARKE

There seems to be a terrible misunderstanding on the part of a great many people to the effect that when you cease to believe you may cease to behave.

—LOUIS KRONENBERGER

We have just enough religion to make us hate, but not enough to make us love one another.

—JONATHAN SWIFT

Christian morality (so called) has all the characters of a reaction; it is, in great part, a protest against Paganism. Its ideal is negative rather than positive; passive rather than active; Innocence rather than Nobleness; Abstinence from Evil, rather than energetic Pursuit of Good: in its precepts (as has been well said) "thou shalt not" predominates unduly over "thou shalt."

—John Stuart Mill

Morality is of the highest importance—but for us, not for God.

—Albert Einstein

Wherever morality is based on theology, wherever the right is made dependent on divine authority, the most immoral, unjust, infamous things can be justified and established.

—Ludwig Feuerbach

I don't think the belief in God has very much to do with people's moral quality as people.

—Colin McGinn

Atheism, and the related conviction that we have just one life to live, is the only sure way to regard all our fellow creatures as brothers and sisters.

—Christopher Hitchens

The happiest people I have known have been those who gave themselves no concern about their own souls, but did their uttermost to mitigate the miseries of others.

—Elizabeth Cady Stanton

You can pray for someone even if you don't think God exists.

—Gordon Atkinson

The peak of tolerance is most readily achieved by those who are not burdened by convictions.

—Alexander Chase

I have no choice but to be guilty. . . . God is unthinkable if we are innocent.

—ARCHIBALD MACLEISH

Moral indignation is jealousy with a halo.

—H. G. WELLS

Man's responsibility increases as that of the gods decreases.

—ANDRÉ GIDE

The trouble with religious morality comes not from morality's being inescapably pure, but from religion's being incurably unintelligible.

—BERNARD WILLIAMS

It is only by dispelling the clouds and phantoms of Religion, that we shall discover Truth, Reason, and Morality.

—BARON D'HOLBACH

I believe that our obligation is to make life better because it's our obligation to each other as human beings. Not in relation to eternal rewards and infernal punishments.

—Susan Jacoby

Men become civilized, not in proportion to their willingness to believe, but in proportion to their readiness to doubt.

—H. L. Mencken

There is a story... which is fairly well known, told about when missionaries came to Africa, that they had the Bible and we, the natives, had the land. And then they said, "Let us pray," and we dutifully shut our eyes. And when we opened them, why, they now had the land and we had the Bible.

—Desmond Tutu

We are punished by our sins, not for them.

—Elbert Hubbard

It is wonderful how much time good people spend fighting the devil. If they would only expend the same amount of energy loving their fellow men, the devil would die in his own tracks of ennui.

—Helen Keller

[For St. Francis of Assisi,] religion was not a thing like a theory but a thing like a love-affair.

—G. K. Chesterton

Atheism leaves a man to sense, to philosophy, to natural piety, to laws, to reputation; all which may be guides to an outward moral virtue, though religion were not; but superstition dismounts all these and erecteth an absolute monarchy in the minds of men.

—Francis Bacon

What remains, then, for those who cannot pray. . . . This alone, and this is enough: To love virtue, to love truth.

—John Burroughs

All religions have based morality on obedience, that is to say, on voluntary slavery. That is why they have always been more pernicious than any political organization. For the latter makes use of violence, the former—of the corruption of the will.

—ALEXANDER HERZEN

I don't believe in God. My God is patriotism. Teach a man to be a good citizen and you have solved the problem of life.

—ANDREW CARNEGIE

This is my simple religion. No need for temples. No need for complicated philosophy. Your own mind, your own heart, is the temple; your philosophy is simple kindness.

—THE DALAI LAMA

Science . . . has been accused of undermining morals—but wrongly. The ethical behavior of man is better based on sympathy, education and social relationships, and requires no support from religion. Man's plight would, indeed, be sad if he had to be kept in order through fear of punishment and hope of rewards after death.

—ALBERT EINSTEIN

"Heaven help us!" said the old religion; the new one, from its very lack of that faith, will teach us all the more to help one another.

—GEORGE ELIOT

My only wish is . . . to transform friends of God into friends of man, believers into thinkers, devotees of prayer into devotees of work, candidates for the hereafter into students of this world, Christians who, by their own profession and admission, are *"half animal, half angel,"* into *persons*, into *whole persons*.

—LUDWIG FEUERBACH

I'm an atheist, and that's *it*. I believe there's nothing we can know except that we should be kind to each other and do what we can for other people.

—Katharine Hepburn

He who has made great moral progress ceases to pray.

—Immanuel Kant

I have been astonished that Men could die Martyrs for religion—I have shudder'd at it—I shudder no more—I could be martyr'd for my Religion—Love is my religion—I could die for that.

—John Keats

God give me unclouded eyes and freedom from haste. God give me a quiet and relentless anger against all pretense and all pretentious work and all work left slack and unfinished. God give me a restlessness whereby I may neither sleep nor accept praise till my observed results equal my calculated results or in pious glee I discover and assault my error. God give me strength not to trust God!

—Sinclair Lewis

Strange is our situation here upon earth. Each of us comes for a short visit, not knowing why, yet sometimes seeming to divine a purpose. From the standpoint of daily life, however, there is one thing we do know: that man is here for the sake of other men—above all for those upon whose smile and well-being our own happiness depends, and also for the countless unknown souls with whose fate we are connected by a bond of sympathy.

—ALBERT EINSTEIN

Thus was I forced, through seeing the error of their foundation, to abandon all belief in every religion which had been taught to man. But my religious feelings were immediately replaced by the spirit of universal char-ity—not for a sect or a party, or for a country or a colour, but for the human race, and with a real and ardent desire to do them good.

—ROBERT OWEN

I believe in salvation through economic, social and spiritual freedom.

—ELBERT HUBBARD

Religion is an insult to human dignity. With or without it you would have good people doing good things and evil people doing evil things. But for good people to do evil things, that takes religion.

—STEVEN WEINBERG

Pointing to another world will never stop vice among us; shedding light over this world can alone help us.

—WALT WHITMAN

I believe in the equality of man; and I believe that religious duties consist in doing justice, loving mercy, and endeavoring to make our fellow-creatures happy.

—THOMAS PAINE

Not one man in ten thousand has goodness of heart or strength of mind to be an atheist.

—Samuel Taylor Coleridge

So many gods, so many creeds,
So many paths that wind and wind,
While just the art of being kind
Is all the sad world needs.

—Ella Wheeler Wilcox

xv

BOOK OF KNOWLEDGE

The liberation of the human mind has never been furthered by such learned dunderheads; it has been furthered by gay fellows who heaved dead cats into sanctuaries and then went roistering down the highways of the world, proving to all men that doubt, after all, was safe—that the god in the sanctuary was finite in his power, and hence a fraud.

—H. L. MENCKEN (and GEORGE JEAN NATHAN)

I have, therefore, described religion as the metaphysics of the people.

—ARTHUR SCHOPENHAUER

The more I study religions the more I am convinced that man never worshipped anything but himself.

—Richard Francis Burton

The Atheist does not say "There is no God," but he says, "I know not what you mean by God; I am without idea of God; the word 'God' is to me a sound conveying no clear or distinct affirmation."

—Charles Bradlaugh

God is a holding place for everything we don't understand.

—Betty Sue Flowers

Doubt is not a pleasant condition, but certainty is a ridiculous one.

—Voltaire

What's the difference between the Lone Ranger and God? There really is a Lone Ranger.

—EDWARD ABBEY

Human kind cannot bear very much reality.

—T. S. ELIOT

Of religion I know nothing—at least, in its favor.

—LORD BYRON

Religion has treated knowledge sometimes as an enemy, sometimes as a hostage; often as a captive, and more often as a child.

—CHARLES CALEB COLTON

God, Immortality, Duty . . . how inconceivable was the *first*, how unbelievable the *second*, and yet how peremptory and absolute the *third*.

—GEORGE ELIOT

If ignorance of nature gave birth to gods, knowledge of nature is made for their destruction.

—Percy Bysshe Shelley

Although atheism might have been *logically* tenable before Darwin, Darwin made it possible to be an intellectually fulfilled atheist.

—Richard Dawkins

Fear prophets . . . and those prepared to die for the truth, for as a rule they make many others die with them, often before them, at times instead of them.

—Umberto Eco

I can live with doubt and uncertainty. I think it's much more interesting to live not knowing than to have answers which might be wrong.

—Richard P. Feynman

Truth is tough. It will not break, like a bubble, at a touch; nay, you may kick it about all day, like a football, and it will be round and full at evening.

—Oliver Wendell Holmes

You never see animals going through the absurd and often horrible fooleries of magic and religion. . . . Only man behaves with such gratuitous folly. It is the price he has to pay for being intelligent, but not, as yet, quite intelligent enough.

—Aldous Huxley

I wouldn't want my doctor thinking that intelligent design was an equally plausible hypothesis to evolution any more than I would want my airplane pilot believing in the flat Earth.

—James Langer

Know then thyself, presume not God to scan,
The proper study of Mankind is Man.

—Alexander Pope

For me, it is far better to grasp the Universe as it really is than to persist in delusion, however satisfying and reassuring.

—Carl Sagan

Religions are like glow-worms: they need darkness in order to shine. A certain degree of general ignorance is the condition for the existence of any religion, the element in which alone it is able to live.

—Arthur Schopenhauer

Modest doubt is called the beacon of the wise.

—William Shakespeare

I am against religion because it teaches us to be satisfied with not understanding the world.

—Richard Dawkins

Reality is that which, when you stop believing in it, doesn't go away.*

—Philip K. Dick

Well I don't think we're *for* anything. We're just products of evolution. You can say "Gee, your life must be pretty bleak if you don't think there's a purpose." But I'm anticipating having a good lunch.

—James Watson

What do I know about God and the purpose of life? I know that this world exists.

—Ludwig Wittgenstein

* This was Dick's response, in 1972, to a Canadian college student who was working on a paper for a philosophy class and asked him for a one-sentence definition of reality.

The memory of my own suffering has prevented me from ever shadowing one young soul with any of the superstitions of the Christian religion.

—Elizabeth Cady Stanton

Just think of the tragedy of teaching children not to doubt.

—Clarence Darrow

xvi

BOOK OF TRUTH

For what is Truth? In matters of religion, it is simply the opinion that has survived. In matters of science, it is the ultimate sensation.

—OSCAR WILDE

Religion is fundamentally opposed to everything I hold in veneration—courage, clear thinking, honesty, fairness, and, above all, love of the truth.

—H. L. MENCKEN

It can do truth no service to blink the fact, known to all who have the most ordinary acquaintance with literary history, that a large portion of the noblest and most valuable moral teaching has been the work, not only of men who did not know, but of men who knew and rejected, the Christian faith.

—John Stuart Mill

Truth is more of a stranger than fiction.

—Mark Twain

Most religions have merely canonized a few products of ancient ignorance and derangement and passed them down to us as though they were primordial truths. This leaves billions of us believing what no sane person could believe on his own.

—Sam Harris

The concepts of truth may differ. But all admit and respect truth. That truth I call God. For sometime I was saying, "God is Truth," but that did not satisfy me. So now I say, "Truth is God."

—MOHANDAS K. GANDHI

The truths of religion are never so well understood as by those who have lost the power of reasoning.

—VOLTAIRE

Religion is about turning untested belief into unshakeable truth through the power of institutions and the passage of time.

—RICHARD DAWKINS

If devotion to truth is the hallmark of morality, then there is no greater, nobler, more heroic form of devotion than the act of a man who assumes the responsibility of thinking. . . . The alleged short-cut to knowledge, which is faith, is only a short-circuit destroying the mind.

—AYN RAND

The history of our race, and each individual's experience, are sown thick with evidence that a truth is not hard to kill and that a lie told well is immortal.

—MARK TWAIN

The truth never flaunted a sign.

—EMILY DICKINSON

I am a friend of truth, but no friend at all to martyrdom.

—VOLTAIRE

It is so little true that *martyrs* offer any support to the truth of a cause that I am inclined to deny that any martyr has ever had anything to do with the truth at all.

—FRIEDRICH NIETZSCHE

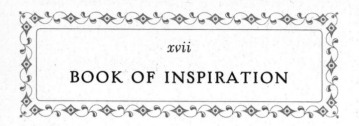

xvii

BOOK OF INSPIRATION

Belief? What do I believe in? I believe in sun. In rock. In the dogma of the sun and the doctrine of the rock. I believe in blood, fire, woman, rivers, eagles, storm, drums, flutes, banjos, and broom-tailed horses.

—Edward Abbey

If I had to choose a religion I think I should become a worshipper of the sun. The sun gives to all things life and fertility. It is the true God of the earth.

—Napoléon Bonaparte

I believe in Michael Angelo, Velasquez, and Rembrandt; in the might of design, the mystery of color, the redemption of all things by Beauty everlasting, and the message of Art that has made these hands blessed.

—GEORGE BERNARD SHAW

I feel no need of any faith than my faith in human beings. Like Confucius of old, I am so absorbed in the wonder of earth and the life upon it that I cannot think of heaven or angels.

—PEARL S. BUCK

Joy in the universe, and keen curiosity about it all—that has been my religion.

—JOHN BURROUGHS

The fairest thing we can experience is the mysterious. It is the fundamental emotion which stands at the cradle of true art and true science.

—ALBERT EINSTEIN

The church of this country is not only indifferent to the wrongs of the slave, it actually takes sides with the oppressors. . . . For my part, I would say, welcome infidelity! welcome atheism! welcome anything! in preference to the gospel, *as preached by those Divines!* They convert the very name of religion into an engine of tyranny, and barbarous cruelty, and serve to confirm more infidels, in this age, than all the infidel writings of Thomas Paine, Voltaire, and Bolingbroke, put together, have done!

—Frederick Douglass

Nothing is at last sacred but the integrity of our own mind.

—Ralph Waldo Emerson

Secularism . . . has no mysteries, no mummeries, no priests, no ceremonies, no falsehoods, no miracles, and no persecutions. It considers the lilies of the field, and takes thought for the morrow. It says to the whole world, Work that you may eat, drink, and be clothed; work that you may enjoy; work that you may not want; work that you may give and never need.

—Robert G. Ingersoll

It was, of course, a lie what you read about my religious convictions, a lie which is being systematically repeated. I do not believe in a personal God and I have never denied this but have expressed it clearly. If something is in me which can be called religious then it is the unbounded admiration for the structure of the world so far as our science can reveal it.

—ALBERT EINSTEIN

Like many people, I have no religion, and I am just sitting in a small boat drifting with the tide. I live in the doubts of my duty. . . . I think there is dignity in this, just to go on working.

—FEDERICO FELLINI

Atheism in its negation of gods is at the same time the strongest affirmation of man, and through man, the eternal yea to life, purpose, and beauty.

—EMMA GOLDMAN

Let us use our energy and our initiative to solve our problems without relying on prayers and wishful thinking. When we have faith in ourselves, we will find we do not have to have faith in gods.

—Ruth Hurmence Green

I would do anything—light candles, say chants, recite prayers, give alms—on the off-chance that one of them would work.

—Ann Richards

My Imagination is a Monastery and I am its Monk.

—John Keats

There is no God any more divine than Yourself.

—Walt Whitman

I believe in God, only I spell it Nature.

—Frank Lloyd Wright

The natural is so awesome that we need not go beyond it.

—RUTH HURMENCE GREEN

I can very well do without God both in my life and in my painting, but I cannot, ill as I am, do without something which is greater than I, which is my life—the power to create.

—VINCENT VAN GOGH

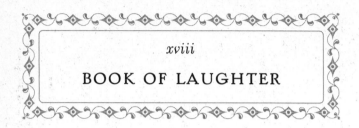

xviii

BOOK OF LAUGHTER

CREATOR—A comedian whose audience is afraid to laugh.

—H. L. MENCKEN

If only God would give me some clear sign! Like making a large deposit in my name at a Swiss bank.

—WOODY ALLEN

If Jesus had been killed twenty years ago, Catholic school children would be wearing little electric chairs around their necks instead of crosses.

—LENNY BRUCE

Good God, how much reverence can you have for a Supreme Being who finds it necessary to include such phenomena as phlegm and tooth decay in His divine system of creation?

—JOSEPH HELLER

No man of any humor ever founded a religion.

—ROBERT G. INGERSOLL

Forgive, O Lord, my little jokes on Thee
And I'll forgive Thy great big one on me.

—ROBERT FROST

I once wanted to become an atheist. I gave up the idea. They have no holidays.

—HENNY YOUNGMAN

As long as there are tests, there will be prayer in schools.

—ANONYMOUS

I'm not normally a praying man, but if you're up there, please save me, Superman.

—Homer Simpson

Why should we take advice on sex from the Pope? If he knows anything about it, he shouldn't.

—George Bernard Shaw

I do not believe in God. I believe in cashmere.

—Fran Lebowitz

God is love, but get it in writing.

—Gypsy Rose Lee

CHRISTIAN: One who believes that the New Testament is a divinely inspired book admirably suited to the spiritual needs of his neighbor.

—AMBROSE BIERCE

The total absence of humor from the Bible is one of the most singular things in all literature.

—ALFRED NORTH WHITEHEAD

If Jesus was Jewish, how come he has a Mexican name?

—ANONYMOUS

Every man thinks God is on his side. The rich and powerful know He is.

—JEAN ANOUILH

There are three religious truths:

1. Jews do not recognize Jesus as the Messiah.
2. Protestants do not recognize the Pope as the leader of the Christian faith.
3. Baptists do not recognize each other in the liquor store or at Hooters.

—ANONYMOUS

Man is a dog's ideal of what God should be.

—HOLBROOK JACKSON

But why do born-again people so often make you wish they'd never been born the first time?

—KATHARINE WHITEHORN

An atheist is a guy who watches a Notre Dame–SMU football game and doesn't care who wins.

—DWIGHT D. EISENHOWER

Orthodoxy is my doxy—heterodoxy is another man's doxy.

—William Warburton

Nah, there's no bigger atheist than me. Well, I take that back. I'm a cancer screening away from going agnostic and a biopsy away from full-fledged Christian.

—Adam Carolla

When suffering comes, we yearn for some sign from God, forgetting we have just had one.

—Mignon McLaughlin

I want to play the role of Jesus. I'm a logical choice. I look the part. I'm a Jew. And I'm a comedian. . . . And I'm an atheist, so I'd be able to look at the character objectively. Who else could do that?

—Charlie Chaplin

God wanted to have a holiday, so He asked St. Peter for suggestions on where to go.

"Why not go to Jupiter?" asked St. Peter.

"No, too much gravity, too much stomping around," said God.

"Well, how about Mercury?"

"No, it's too hot there."

"Okay," said St. Peter, "what about Earth?"

"No," said God, "They're such horrible gossips. When I was there 2000 years ago, I had an affair with a Jewish woman, and they're still talking about it."

—Anonymous

It's hard to be religious when certain people are never incinerated by bolts of lightning.

—Bill Watterson

The atheist, by merely being in touch with reality, appears shamefully out of touch with the fantasy life of his neighbors.

—Sam Harris

When I was a kid I used to pray every night for a new bicycle. Then I realized that the Lord doesn't work that way. So I stole one and asked Him to forgive me.

—Emo Philips

Dear Lord, the gods have been good to me, and I am thankful. For the first time in my life everything is absolutely perfect just the way it is. So here's the deal. You freeze everything as it is, and I won't ask for anything more. If that is okay, please give me absolutely no sign. Okay, deal. In gratitude, I present you this offering of milk and cookies. If you want me to eat them for you, give me no sign. That will be done (*gulp, gulp, munch, munch, munch*).

—Homer Simpson

We must respect the other fellow's religion, but only in the sense and to the extent that we respect his theory that his wife is beautiful and his children smart.

—H. L. MENCKEN

What if everything is an illusion and nothing exists? In that case, I definitely overpaid for my carpet.

—WOODY ALLEN

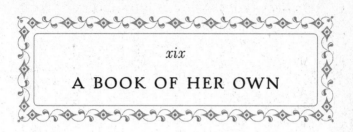

So long as [women themselves] mistake superstition for religious revelation, they will be content with the position and opportunities assigned them by scholastic theology. . . . Their religious nature is warped and twisted through generations . . . which fact, by the way, is the greatest stumbling block in the path of equal suffrage to-day.

—Elizabeth Cady Stanton

I am influenced in my own conduct at the present time by far higher considerations, and by a nobler idea of duty, than I ever was while I held the evangelical beliefs.

—George Eliot

I do not believe in God, because I believe in man. Whatever his mistakes, man has for thousands of years past been working to undo the botched job your God has made.

—EMMA GOLDMAN

An omnipotent, all-knowing tyrant is not so different from earthly dictators who made everything and everybody mere cogs in the machine which they controlled. An atheism that rejects such a God is amply justified.

—KAREN ARMSTRONG

All religions are the same: religion is basically guilt, with different holidays.

—CATHY LADMAN

I distrust those people who know so well what God wants them to do, because I notice it always coincides with their own desires.

—SUSAN B. ANTHONY

Some keep the Sabbath going to church;
I keep it staying at home,
With a bobolink for a chorister,
And an orchard for a dome.

—EMILY DICKINSON

The bible teaches that a father may sell his daughter for a slave (Ex. xxi, 7), that he may sacrifice her purity to a mob (Judges xix, 24), and that he may murder her, and still be a good father and a holy man. It teaches that a man may have any number of wives; that he may sell them, give them away, or change them around, and still be a perfect gentleman, a good husband, a righteous man, and one of God's most intimate friends; and that is a pretty good position for a beginning.

—HELEN H. GARDNER

Among all forms of mistake, prophecy is the most gratuitous.

—GEORGE ELIOT

She was a good Christian woman with a large respect for religion, though she did not, of course, believe any of it was true.

—FLANNERY O'CONNOR

I find it interesting that the meanest life, the poorest existence, is attributed to God's will, but as human beings become more affluent, as their living standard and style begin to ascend the material scale, God descends the scale of responsibility at a commensurate speed.

—MAYA ANGELOU

Religion is probably, after sex, the second oldest resource which human beings have available to them for blowing their minds.

—SUSAN SONTAG

By the year 2000, we will, I hope, raise our children to believe in human potential, not God.

—GLORIA STEINEM

I have endeavoured to dissipate these religious superstitions from the minds of women, and base their faith on science and reason, where I found for myself at last that peace and comfort I could never find in the Bible and the church.

—ELIZABETH CADY STANTON

The philosophy of Atheism represents a concept of life without any metaphysical Beyond or Divine Regulator. It is the concept of an actual, real world with its liberating, expanding and beautifying possibilities, as against an unreal world, which, with its spirits, oracles, and mean contentment has kept humanity in helpless degradation.

—EMMA GOLDMAN

There ain't no answer. There ain't going to be any answer. There never has been an answer. That's the answer.

—GERTRUDE STEIN

Back in my early childhood, I had learned that God doesn't fight on any army's side. So there was little point in praying. Nonetheless, before every battle prayers were read, all kinds of incantations were recited, staged by all sorts of preachers. We attended these ceremonies, and I saw how the soldiers stood in place, as though they couldn't believe their ears. . . . I couldn't believe it either, but I counted for nothing. . . . Since then, I've given up any belief in God, in a "light" that leads us, or anything of this sort. Goethe has said, "If God created this world, then he should review his plan."

—MARLENE DIETRICH

No Gods, No Masters.*

—MARGARET SANGER

* "No Gods, No Masters" was printed on the masthead of Margaret Sanger's feminist journal *The Woman Rebel*. This slogan originally appeared on a sign carried by Industrial Workers of the World (IWW) union members at the strike in Lawrence, Massachusetts, in 1912. The sign read: "Arise!!! Slaves of the World!!! No God! No Master! One for all and all for one."

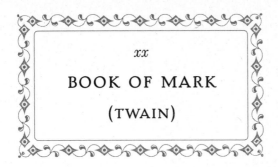

xx

BOOK OF MARK

(TWAIN)

Faith is believing what you know ain't so.

Most people are bothered by those passages in Scripture which they cannot understand; but as for me, I always noticed that the passages in Scripture which trouble me most are those that I do understand.

Religion consists in a set of things which the average man thinks he believes, and wishes he was certain.

But who prays for Satan? Who, in eighteen centuries, has had the common humanity to pray for the one sinner that needed it most?

᠕᠊ᢒ᠊ᢲ

In religion and politics people's beliefs and convictions are in almost every case gotten at second-hand, and without examination, from authorities who have not themselves examined the questions at issue but have taken them at second-hand from other non-examiners, whose opinions about them were not worth a brass farthing.

᠕᠊ᢒ᠊ᢲ

It is best to read the weather forecast before we pray for rain.

᠕᠊ᢒ᠊ᢲ

Ours is a terrible religion. The fleets of the world could swim in spacious comfort in the innocent blood it has spilt.

᠕᠊ᢒ᠊ᢲ

Blasphemy? No, it is not blasphemy. If God is as vast as that, He is above blasphemy; if He is as little as that, He is beneath it.

⊱⧉⊰

Man is a marvelous curiosity. . . . He thinks he is the Creator's pet. . . . He even believes the Creator loves him; has a passion for him; sits up nights to admire him; yes, and watch over him and keep him out of trouble. He prays to Him, and thinks He listens. Isn't it a quaint idea?

⊱⧉⊰

If Christ were here now, there is one thing he would *not* be—a Christian.

⊱⧉⊰

Man is the Religious Animal. . . . He is the only animal that has the True Religion—several of them. He is the only animal that loves his neighbor as himself, and cuts his throat if his theology isn't straight. He has made a graveyard of the globe in trying his honest best to smooth his brother's path to happiness and heaven.

⊱⧉⊰

The church is always trying to get other people to reform; it might not be a bad idea to reform itself a little, by way of example.

❧

One of the proofs of the immortality of the soul is that myriads have believed it. They also believed the world was flat.

❧

Nothing exists; all is a dream. God—man—the world—the sun, the moon, the wilderness of stars—a dream, all a dream; they have no existence. *Nothing exists save empty space—and you!*

❧

[The Bible] has noble poetry in it . . . and some good morals; and some execrable morals; and a wealth of obscenity; and upwards of a thousand lies.

❧

Many of these people have the reasoning faculty, but no one uses it in religious matters.

If the cholera or black plague should come to these shores, perhaps the bulk of the nation would pray to be delivered from it, but the rest would put their trust in the Health Board of the City of New York.

❧

What God lacks is convictions—stability of character. He ought to be a *Presbyterian* or a *Catholic* or *something*—not try to be everything.

❧

The best minds will tell you that when a man has begotten a child he is morally bound to tenderly care for it, protect it from hurt, shield it from disease, clothe it, feed it, bear with its waywardness, lay no hand upon it save in kindness and for its own good, and never in any case inflict upon it a wanton cruelty. God's treatment of his earthly children, every day and every night, is the exact opposite of all that, yet those best minds warmly justify these crimes, condone them, excuse them, and indignantly refuse to regard them as crimes at all, when *he* commits them.

❧

Nothing agrees with me. If I drink coffee it gives me dispepsia; if I drink wine it gives me the gout; if I go to church it gives me dysentery.

꠵

A man is accepted into a church for what he believes and he is turned out for what he knows.

꠵

[Go to] heaven for climate, and hell for society.

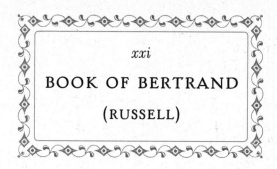

xxi

BOOK OF BERTRAND

(RUSSELL)

No man treats a motorcar as foolishly as he treats an-other human being. When the car will not go, he does not attribute its annoying behavior to sin; he does not say, "You are a wicked motorcar, and I shall not give you any more petrol until you go." He attempts to find out what is wrong and set it right.

One is often told that it is a very wrong thing to attack religion, because religion makes men virtuous. So I am told; I have not noticed it.

If I were granted omnipotence, and millions of years to experiment in, I should not think Man much to boast of as the final result of all my efforts.

⊱✦⊰

The fundamental cause of the trouble is that in the modern world the stupid are cocksure while the intelligent are full of doubt.

⊱✦⊰

So far as I can remember, there is not one word in the Gospels in praise of intelligence.

⊱✦⊰

I am suggesting that, in all our beliefs, we should admit a "probable error." If you believe (say) that democracy is better than fascism, you should still admit the possibility of error, though the possibility may be very small. The possibility may be so small that you are willing to kill and die for your belief, yet the knowledge that the possibility exists may keep you from advocating large-scale persecutions and cruelties such as are almost always practiced by those who admit no doubts.

⊱✦⊰

To teach how to live without certainty, and yet without being paralyzed by hesitation, is perhaps the chief thing that philosophy, in our age, can still do for those who study it.

❧

I believe that when I die I shall rot, and nothing of my ego will survive. I am not young, and I love life. But I should scorn to shiver with terror at the thought of annihilation. Happiness is nonetheless true happiness because it must come to an end, nor do thought and love lose their value because they are not everlasting.*

❧

And if there were a God, I think it very unlikely that He would have such an uneasy vanity as to be offended by those who doubt His existence.

❧

* Originally published in 1925, in a small book titled *What I Believe*. That same book was used as a piece of evidence in 1940, when the New York City court system deemed Russell morally unfit to teach at City College.

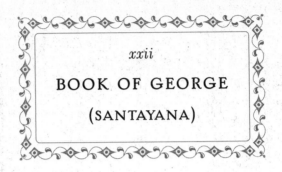

xxii

BOOK OF GEORGE

(SANTAYANA)

Prayer, among sane people, has never superseded practical efforts to secure the desired end.

I can always say to myself that my atheism, like that of Spinoza, is true piety towards the universe and denies only gods fashioned by men in their own image, to be servants of their human interests.

Fanaticism consists in redoubling your effort when you have forgotten your aim.

Religions are the great fairy-tales of the conscience.

⋙⋘

That fear first created the gods is perhaps as true as anything so brief could be on so great a subject.

⋙⋘

Each religion, so dear to those whose life it sanctifies, and fulfilling so necessary a function in the society that has adopted it, necessarily contradicts every other religion, and probably contradicts itself.

⋙⋘

The idea of Christ is much older than Christianity.

⋙⋘

Religion is the natural reaction of the imagination when confronted by the difficulties of a truculent world.

⋙⋘

Faith in the supernatural is a desperate wager made by man at the lowest ebb of his fortunes.

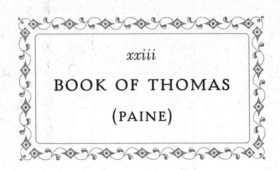

I do not believe in the creed professed by the Jewish Church, by the Roman Church, by the Greek Church, by the Turkish Church, by the Protestant Church, nor by any church that I know of. My own mind is my own church.

≈✦≈

One good schoolmaster is of more use than a hundred priests.

To argue with a man who has renounced the use and authority of reason, and whose philosophy consists in holding humanity in contempt, is like administering medicine to the dead, or endeavoring to convert an atheist by scripture.

❧❧❧

No falsehood is so fatal as that which is made an article of faith.

❧❧❧

Accustom a people to believe that priests, or any other class of men, can forgive sins, and you will have sins in abundance.

❧❧❧

The declaration which says that God *visits the sins of the fathers upon the children* . . . is contrary to every principle of moral justice.

❧❧❧

Whenever we read the obscene stories, the voluptuous debaucheries, the cruel and torturous executions, the unrelenting vindictiveness, with which more than half the Bible is filled, it would be more consistent that we called it the word of a demon, than the Word of God. It is a history of wickedness, that has served to corrupt and brutalize mankind; and, for my own part, I sincerely detest it, as I detest everything that is cruel.

❧

What is it the Bible teaches us?—rapine, cruelty, and murder. What is it the Testament teaches us?—to believe that the Almighty committed debauchery with a woman engaged to be married, and the belief of this debauchery is called faith.

❧

All the tales of miracles with which the Old and New Testament are filled, are fit only for impostors to preach and fools to believe.

❧

Of all the tyrannies that afflict mankind, tyranny in religion is the worst: every other species of tyranny is limited to the world we live in; but this attempts a stride beyond the grave, and seeks to pursue us into eternity.

❧❧

The story of Jesus Christ appearing after he was dead is the story of an apparition, such as timid imaginations can always create in vision, and credulity believe.

❧❧

All national institutions of churches, whether Jewish, Christian or Turkish, appear to me no other than human inventions, set up to terrify and enslave mankind, and monopolize power and profit.

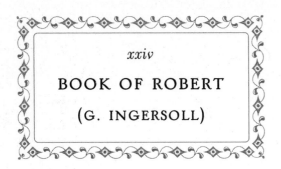

xxiv

BOOK OF ROBERT

(G. INGERSOLL)

If a man would follow, today, the teachings of the Old Testament he would be a criminal. If he would strictly follow the teachings of the New, he would be insane.

The inventor of the plow did more good than the maker of the first rosary—because, say what you will, plowing is better than praying.

With soap, baptism is a good thing.

In Nature there are neither rewards nor punishments—
there are consequences.

Yes; if a man really believes that God once upheld slavery; that he commanded soldiers to kill women and
babes; that he believed in polygamy; that he persecuted
for opinion's sake; that he will punish forever, and that
he hates an unbeliever, the effect in my judgment will be
bad. It always has been bad. This belief built the dun-
geons of the Inquisition. This belief made the Puritan
murder the Quaker.

Few nations have been so poor as to have but one god.
Gods were made so easily, and the raw material cost so
little, that generally the god market was fairly glutted,
and heaven crammed with these phantoms.

If the founder of Christianity had plainly said: "It is not necessary to believe in order to be saved; it is only necessary to do, and he who really loves his fellow-men, who is kind, honest, just and charitable, is to be forever blest"—if he had only said that, there would probably have been but little persecution.

The old doctrine that God ... rewarded the virtuous and punished the wicked is gradually fading from the mind. We know that some of the worst men have what the world calls success. We know that some of the best men lie upon the straw of failure. We know that honesty goes hungry, while larceny sits at the banquet. We know that the vicious have every physical comfort, while the virtuous are often clad in rags.

Fear believes—courage doubts. Fear falls upon the earth and prays—courage stands erect and thinks. Fear retreats—courage advances. Fear is barbarism—courage is civilization. Fear believes in witchcraft, in devils and in ghosts. Fear is religion, courage is science.

The hands that help are better far than the lips that pray.

⋙⋘

The churches have no confidence in each other. Why? Because they are acquainted with each other.

⋙⋘

The notion that faith in Christ is to be rewarded by an eternity of bliss, while a dependence upon reason, observation, and experience merits everlasting pain, is too absurd for refutation, and can be relieved only by that unhappy mixture of insanity and ignorance, called "faith."

⋙⋘

Labor is the only prayer that Nature answers; it is the only prayer that deserves an answer—good, honest, noble work.

⋙⋘

The inspiration of the Bible depends upon the ignorance of him who reads.

It may be that ministers really think that their prayers do good and it may be that frogs imagine that their croaking brings spring.

❧

The clergy know, I know, that they know that they do not know.

❧

Christianity has such a contemptible opinion of human nature that it does not believe a man can tell the truth unless frightened by a belief in God. No lower opinion of the human race has ever been expressed.

❧

I cannot see why we should expect an infinite God to do better in another world than he does in this.

❧

Honest investigation is utterly impossible within the pale of any church, for the reason, that if you think the church is right you will not investigate, and if you think it wrong, the church will investigate you.

As long as every question is answered by the word "god," scientific inquiry is simply impossible.

I have little confidence in any enterprise or business or investment that promises dividends only after the death of the stockholders.

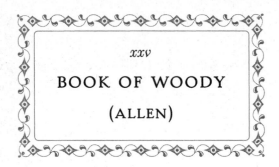

Well, I believe that there's somebody out there who watches over us. Unfortunately, it's the government.

✦

You know, if it turns out that there is a God, I don't think that he's evil. I think the worst you can say about him is that basically he's an underachiever.

✦

I was thrown out of NYU my freshman year for cheating on my metaphysics final. You know, I looked within the soul of the boy sitting next to me.

✦

To you I'm an atheist. . . . To God I'm the loyal opposition.

Eternal nothingness is O.K. if you're dressed for it.

I don't believe in an afterlife, although I am bringing a change of underwear.

I don't want to achieve immortality through my work. I want to achieve it through not dying.

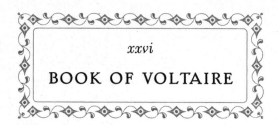

Sect and error are synonymous terms.

※<

The atheist preserves his reason, which checks his propensity to mischief, while the fanatic is under the influence of a madness which is constantly urging him on.

※<

There is no sect in geometry; one does not refer to a Euclidean or Archimedean.

※<

[Theological religion] is the source of all imaginable follies and disorders; it is the mother of fanaticism and civil discord; it is the enemy of mankind.

❧

I always said a very short prayer to God; here it is: "My God! make my enemies very ridiculous!" God heard my prayer.

❧

If God created us in his own image, we have more than reciprocated.

❧

Nothing can be more contrary to religion and the clergy than reason and common sense.

❧

If we believe absurdities, we shall commit atrocities.

❧

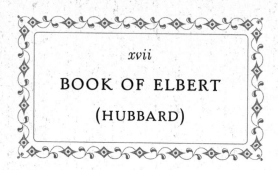

xvii

BOOK OF ELBERT

(HUBBARD)

DOGMA: A lie imperiously reiterated and authoritatively injected into the mind of one or more persons who believe they believe what some one else believes.

Theology is an attempt to explain a subject by men who do not understand it. The intent is not to tell the truth but to satisfy the questioner.

GOD: 1. The John Doe of philosophy and religion. 2. The first atheist.

⋙⋘

A creed is an ossified metaphor.

⋙⋘

Formal religion was organized for slaves: it offered them consolation which earth did not provide.

⋙⋘

And that what we call God's justice is only man's idea of what he would do if he were God.

⋙⋘

I believe in sunshine, fresh air, friendship, calm sleep, beautiful thoughts.

Theology, by diverting the attention of men from this life to another, and by endeavoring to coerce all men into one religion, constantly preaching that this world is full of misery but the next world would be beautiful—or not, as the case may be—has forced on men the thought of fear, where otherwise there might have been the happy abandon of Nature.

Give us a Religion that will help us live—we can die without assistance.

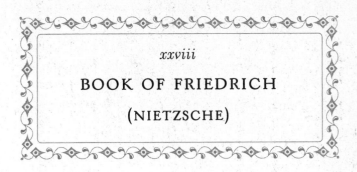

xxviii

BOOK OF FRIEDRICH

(NIETZSCHE)

The Christian resolve to find the world ugly and bad has made the world ugly and bad.

❧

God is a thought—it maketh all the straight crooked, and all that standeth reel.

❧

[Stendhal] robbed me of the best atheist joke which precisely I could have made: "God's only excuse is that he does not exist."

❧

Better songs would they have to sing, for me to believe in their Savior: more! like saved ones would his disciples have to appear unto me!

❧

Mystical explanations are considered deep. The truth is that they are not even superficial.

❧

Two great European narcotics, alcohol and Christianity.

❧

[Jesus] died too early; he himself would have disavowed his doctrine had he attained to my age!

❧

The fact that faith may in certain circumstances save, the fact that salvation as the result of an *idée fixe* does not constitute a true idea, the fact that faith moves *no* mountains, but may very readily raise them where previously they did not exist—all these things are made sufficiently clear by a mere casual stroll through a *lunatic asylum*.

❧

God is dead; but given the way of men, there may still be caves for thousands of years in which his shadow will be shown.

⋊⋉

What? Is man just one of God's mistakes? Or is God just one of man's?

⋊⋉

The spiritualization of sensuality is called *love*: it is a great triumph over Christianity.

⋊⋉

"Faith" means the will to avoid knowing what is true.

⋊⋉

There is not enough love and goodness in the world for us to be permitted to give any of it away to imaginary things.

⋊⋉

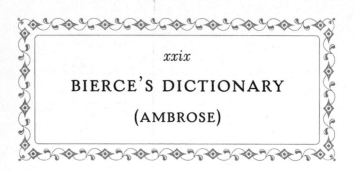

xxix

BIERCE'S DICTIONARY

(AMBROSE)

CLERGYMAN, *n*. A man who undertakes the management of our spiritual affairs as a method of bettering his temporal ones.

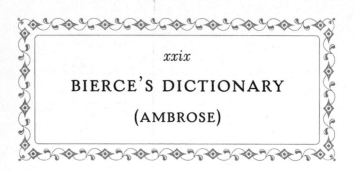

FAITH, *n*. Belief without evidence in what is told by one who speaks without knowledge, of things without parallel.

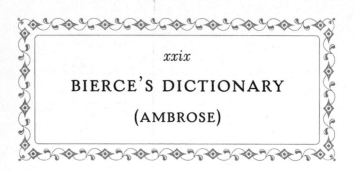

HEATHEN, *n*. A benighted creature who has the folly to worship something that he can see and feel.

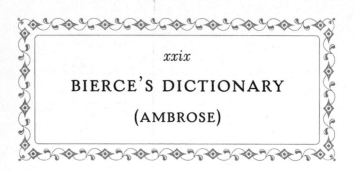

IMPIETY, *n.* Your irreverence toward my deity.

✦

INFIDEL, *n.* In New York, one who does not believe in the Christian religion; in Constantinople, one who does.

✦

PRAY, *v.* To ask that the laws of the universe be annulled in behalf of a single petitioner confessedly unworthy.

✦

REDEMPTION, *n.* Deliverance of sinners from the penalty of their sin, through their murder of the deity against whom they sinned.

✦

RELIGION, *n.* A daughter of Hope and Fear, explaining to Ignorance the nature of the Unknowable.

✦

REVERENCE, *n.* The spiritual attitude of a man to a god and a dog to a man.

SAINT, *n*. A dead sinner revised and edited.

⇒⧫⇐

SCRIPTURES, *n*. The sacred books of our holy religion, as distinguished from the false and profane writings on which all other faiths are based.

Men are most apt to believe what they least understand.
— Michel Eyquem de Montaigne

If I were not an atheist, I would believe in a God who would choose to save people on the basis of the totality of their lives and not the pattern of their words. I think he would prefer an honest and righteous atheist to a TV preacher whose every word is God, God, God, and whose every deed is foul, foul, foul.

— Isaac Asimov

If someone were to prove to me—right this minute—that God, in all his luminousness, exists, it wouldn't change a single aspect of my behavior.

—Luis Buñuel

If I am a fool, it is, at least, a doubting one; and I envy no one the certainty of his self-approved wisdom.

—Lord Byron

I have been into many of the ancient cathedrals—grand, wonderful, mysterious. But I always leave them with a feeling of indignation because of the generations of human beings who have struggled in poverty to build these altars to an unknown god.

—Elizabeth Cady Stanton

I give money for church organs in the hope that the organ music will distract the congregation's attention from the rest of the service.

—Andrew Carnegie

We have men sold to build churches, women sold to support the gospel, and babes sold to purchase Bibles for the *poor heathen! all for the glory of God and the good of soul!* The slave auctioneer's bell and the church-going bell chime in with each other, and the bitter cries of the heart-broken slave are drowned in the religious shouts of his pious master. Revivals of religion and revivals in the slave trade go hand in hand together.

—Frederick Douglass

I cannot imagine a God who rewards and punishes the objects of his creation, whose purposes are modeled after our own—a God, in short, who is but a reflection of human frailty.

—Albert Einstein

Childhood . . . is full of deep sorrows, the meaning of which is unknown. Witness colic and whooping-cough and dread of ghosts, to say nothing of hell and Satan, and an offended Deity in the sky, who was angry when I wanted too much plum-cake.

—George Eliot

My earlier views of the unsoundness of the Christian scheme of salvation and the human origin of the scriptures have become clearer and stronger with advancing years, and I see no reason for thinking I shall ever change them.

—ABRAHAM LINCOLN

Difference in Opinions hath cost many Millions of Lives: For instance, whether *Flesh* be *Bread*, or *Bread* be *Flesh*; whether the Juice of a certain *Berry* be *Blood* or *Wine*.

—JONATHAN SWIFT

Do not pass by my epitaph, traveler.
But having stopped, listen and learn, then go your way.
There is no boat in Hades, no ferryman Charon,
No caretaker Aiakos, no dog Cerberus.
All we who are dead below
Have become bones and ashes, but nothing else.
I have spoken to you honestly, go on, traveler,
Lest even while dead I seem loquacious to you.

—ANCIENT ROMAN TOMBSTONE

What do I know of man's destiny? I could tell you more
about radishes.

—SAMUEL BECKETT

I will have nothing to do with your immortality; we are miserable enough in this life, without the absurdity of speculating upon another.

—Lord Byron

In order to exist just once in the world, it is necessary never again to exist.

—Albert Camus

I feel as I always have, that the earth is the home and the only home of man, and I am convinced that whatever he is to get out of his existence he must get while he is here.

—Clarence Darrow

The skeptic has no illusions about life, nor a vain belief in the promise of immortality. Since this life here and now is all we can know, our most reasonable option is to live it fully.

—Paul Kurtz

Does not Eternity appear dreadful to you. I often get thinking of it and it seems so dark to me that I almost wish there was no Eternity. To think that we must forever live and never cease to be. It seems as if Death which all so dread because it launches us upon an unknown world would be a relief to so endless a state of existence.

—EMILY DICKINSON

We try hard in science to stamp out the influence of wishful thinking, whereas so much of religious thought seems to be nothing else: "I must believe in the afterlife because how could I face it if my life was going to terminate at death?"

—STEVEN WEINBERG

Peacefully they will die . . . and beyond the grave they will find nothing but death. But we shall keep the secret, and for their happiness we shall allure them with the reward of heaven and eternity.

—FYODOR DOSTOYEVSKY

I believe that life should be lived so vividly and so intensely that thoughts of another life, or of a longer life, are not necessary.

—MARJORY STONEMAN DOUGLAS

Assure a man that he has a soul and then frighten him with old wives' tales as to what is to become of it afterwards, and you have a hooked fish, a mental slave!

—THEODORE DREISER

All are inclined to believe what they covet, from a lottery-ticket up to a passport to Paradise; in which, from description, I see nothing very tempting.

—LORD BYRON

Millions long for immortality who don't know what to do with themselves on a rainy Sunday afternoon.

—SUSAN ERTZ

The one inalienable right is to go to destruction in your own way.

—Robert Frost

HEAVEN: The Coney Island of the Christian imagination.

—Elbert Hubbard

I am a hopeless materialist. I see the soul as nothing else than the sum of activities of the organism plus personal habits—plus inherited habits, memories, experiences, of the organism. I believe that when I am dead, I am dead. I believe that with my death I am just as much obliterated as the last mosquito you and I squashed.

—Jack London

Personally, I should not care for immortality in the least. There is nothing better than oblivion, since in oblivion there is no wish unfulfilled. We had it before we were born, yet did not complain. Shall we then whine because we know it will return?

—H. P. Lovecraft

I see life as a dance. Does a dance have to have a meaning? You're dancing because you enjoy it.

—Jackie Mason

My hereafter is here. I am where I'm going, for I am mulch. It's a great comfort to know that in my mulchhood I may nourish a row of parsnips.

—Frank McCourt

From my rotting body flowers shall grow, and I shall be in them.

—Edvard Munch

What time has been wasted during man's destiny in the struggle to decide what man's next world will be like! The keener the effort to find out, the less he knew about the present one he lived in.

—SEAN O'CASEY

Ask yourself whether the dream of heaven and greatness should be left waiting for us in our graves—or whether it should be ours here and now and on this earth.

—AYN RAND

The world is so exquisite, with so much love and moral depth, that there is no reason to deceive ourselves with pretty stories for which there's little good evidence. Far better, it seems to me, in our vulnerability, is to look Death in the eye and to be grateful every day for the brief but magnificent opportunity that life provides.

—CARL SAGAN

To work hard, to live hard, to die hard, and then to go to hell after all would be too damned hard.

—CARL SANDBURG

Heaven, as conventionally conceived, is a place so inane, so dull, so useless, so miserable, that nobody has ever ventured to describe a whole day in heaven, though plenty of people have described a day at the seaside.

—GEORGE BERNARD SHAW

It's an incredible con job, when you think of it, to believe something now in exchange for life after death. Even corporations, with all their reward systems, don't try to make it posthumous.

—GLORIA STEINEM

Life is an opportunity, and it is pregnant with meanings. But the life that you live depends upon your choice.

—PAUL KURTZ

The idea of a good society is something you do not need a religion and eternal punishment to buttress; you need a religion if you are terrified of death.

—GORE VIDAL

It is a curious thing . . . that every creed promises a paradise which will be absolutely uninhabitable for anyone of civilized taste.

—EVELYN WAUGH

Certainty about the next life is simply incompatible with tolerance in this one.

—SAM HARRIS

That it will never come again
Is what makes life so sweet.
—EMILY DICKINSON

Listen, people!
Life is a giant, invisible scale with two sides:
Good and Bad.
You and your beliefs
Are the weights.
The things you do each day
Determine the balance.
Your conscience is a flawless
Judge and jury.
The only question is what you want.

I'm tellin' you the natural facts
For what it's worth.
Listen to me, people.
You make your own heaven and hell
Right here on earth.
    —THE TEMPTATIONS,
        "You Make Your Own Heaven and Hell Right Here"

xxxii

APOCALYPTUS *

I don't know that atheists should be regarded as citizens, nor should they be regarded as patriotic. This is one nation under God.

—George H. W. Bush

* "Apocalyptus" is an invented word derived from the biblical word "apocalypse." Apocalyptus is a weed of bias and prejudice that sprouts in the garden of free speech.

xxxiii

EXODUS

Here lies an atheist
All dressed up
And no place to go.
       —Epitaph

ACKNOWLEDGMENTS

I would like to thank Sara Bader and Liel Liebovitz for their remarkable research and for their sustaining enthusiasm and belief in this project. I am deeply grateful to Robert R. Worth for his suggestions and his keen editorial eye. Thanks, also, to Patty Romeu, Jen Banbury, Josie Peltz, Diane Garvey Nesin and Christina Lowery for their diligent detective work.

INDEX

ABBEY, EDWARD (1927–1989), U.S. environmentalist and writer: 4, 7, 94, 104

ADAMS, DOUGLAS (1952–2001), English science fiction writer: 33, 56

ADAMS, JOHN (1735–1826), second president of the United States: 22

ADAMS, PHILLIP (b. 1939), Australian writer, filmmaker, and radio personality: 11

ALBACETE, MONSIGNOR LORENZO (b. ca. 1941), U.S. theologian and writer: 64

ALLEN, WOODY (Allen Stewart Konigsberg) (b. 1935), U.S. comedian, filmmaker and writer: 110, 118, 146–47

ANGELOU, MAYA (Marguerite Johnson) (b. 1928), U.S. poet and writer: 122

ANOUILH, JEAN (1910–1987), French playwright: 113

ANTHONY, SUSAN B. (1820–1906), U.S. women's rights activist: 120

AQUINAS, SAINT THOMAS (ca. 1225–1274), Italian philosopher and theologian: 50

ARISTOTLE (384–322 B.C.), Greek philosopher: 4, 27

ARMSTRONG, KAREN (b. 1944), English writer and former nun: 120

CAMUS, ALBERT (1913–1960), Algerian-born French writer: 36, 64, 164

CANETTI, ELIAS (1905–1994), Bulgarian-born English writer: 51

CARNEGIE, ANDREW (1835–1919), Scottish-born U.S. industrialist and philanthropist: 86, 160

CAROLLA, ADAM (b. 1964), U.S. comedian: 115

CASTRO, FIDEL (b. 1927), Cuban president: 62

CHAPLIN, CHARLIE (Charles Spencer Chaplin) (1889–1977), English comic actor and film director: 115

CHARRON, PIERRE (1541–1603), French philosopher and theologian: 36

CHASE, ALEXANDER, (b. 1926), U.S. journalist and editor: 82

CHESTERTON, G. K. (1874–1936), English writer: 50, 62, 85

CHOMSKY, NOAM (b. 1928), U.S. linguist and political activist: 21

CICERO, MARCUS TULLIUS (106–43 B.C.), Roman philosopher, orator and statesman: 57

CLARKE, ARTHUR C. (b. 1917), English science-fiction writer: 1, 30, 45, 59, 76, 80

CLEAVER, ELDRIDGE (1935–1998), U.S. civil rights activist: 64

CLIFFORD, W. K. (1845–1879), English mathematician and philosopher: 44

CLOPTON, RICHARD (data unknown): 77

COHEN, CHAPMAN (1868–1954), English writer and atheist activist: 49

COLERIDGE, SAMUEL TAYLOR (1772–1834), English poet, critic, and philosopher: 77, 91

COLTON, CHARLES CALEB (ca. 1780–1832), English author and clergyman: 19, 21, 64, 94

CRICK, FRANCIS (1916–2004), English biophysicist: 46

CRISP, QUENTIN (1908–1999), English writer, performer, and gay rights activist: 36, 55

DAACON, GEORGE (data unknown), Canadian journalist: 57

CREDITS AND PERMISSIONS

Grateful acknowledgment is made for permission to reprint the following quotes:

NOAM CHOMSKY:

Page 21: "Three-quarters of the American population . . ." Noam Chomsky from "Keeping the Rabble in Line," copyright © 1994 by Noam Chomsky and David Barsamian. All rights reserved. Reproduced by special permission.

CHRISTOPHER HITCHENS:

Page 7: "Since it is obviously inconceivable . . ."; p. 29: "Gullibility and credulity are considered . . ."; p. 82: "Atheism, and the related conviction . . ." from Christopher Hitchens, "The Lord and the Intellectuals," copyright © 1982 by *Harpers Magazine*. All rights reserved. Excerpted from the July issue by special permission.

BERTRAND RUSSELL:

Page 31: "What is wanted is not the will . . ." From *Sceptical Essays*, by Bertrand Russell, copyright © 1960, George Allen & Unwin Ltd. Reproduced by permission of Taylor & Francis Books UK and The Bertrand Russell Peace Foundation.

Page 49: "Not to be absolutely certain . . ." From *The Collected Papers of Bertrand Russell: Last Philosophical Testament 1943–68*, Russell, Bertrand, copyright